Please leave your feedback, thoughts, and reviews on Amazon.

See Lieutenant Colonel McBrearty on CBS Los Angeles *Veterans' Voices.*

See Lieutenant Colonel McBrearty on CBS Los Angeles *Sports Central.*

Photo courtesy C.B.S., Los Angeles.

Follow John J. McBrearty's other literary works at:

www.JohnWritesHistory.com

How to Win at Life

A Combat Veteran Shares His Secrets for Success.

by

John J. McBrearty

The views expressed in this publication are those of the author and do not necessarily reflect the official policy or position of Amazon, the Department of War, or the U.S. government.

ISBN: 978-1-966064-23-7 (Paperback)

ISBN: 978-1-966064-24-4 (Hardcover)

Library of Congress Control Number (LCCN): 9781966064213

The cover photo is courtesy of Nicole Rea.

All uncredited photos in this book are the property of John J. McBrearty.

Photos credited to other photographers are their property and appear as a courtesy in this book or are in the public domain.

Published by *Military Books & More Publishing (MBM Publishing).*

Dedication

To my family.

To my fellow veterans.

To my fallen comrades.

"Discovering your purpose in life is essential to living a fulfilling life. "

John J. McBrearty

Veteran, Author, Father & Husband
September 11, 2025

Foreword

A Powerful and Heartfelt Roadmap for Life After Service

LTC John J. McBrearty—known as "Colonel Mack"—offers more than just a book; he shares a mission. Told in a straightforward interview format, *How to Win at Life* provides a sincere and rare look into the heart and mind of a combat veteran who found healing through writing. From coping with grief and PTSD to mentoring young writers and fellow veterans, McBrearty speaks with humility, grace, and raw honesty.

This isn't just for military readers. It's a motivational guide for anyone looking to turn struggle into purpose. McBrearty's voice is genuine and full of faith, and his advice is straightforward, heartfelt, and impactful.

He writes for the soul—for his community, his fallen comrades, and those still fighting silent battles at home. Whether you're a veteran, an aspiring writer, or someone seeking guidance, you'll find comfort and courage in these pages.

To LTC Mack: Keep writing. Your story matters—and so do the stories you inspire others to share.

Derrick C. Darden, Ph.D., Veteran, Academic Professor, and Author

Contents

Lieutenant Colonel, John J. McBrearty, U.S. Army (Retired) interviewed by Nancy, a Ph.D. candidate. (Photo is courtesy of the Author.)

To see the entire interview, go to the following website:

How to Win at Life

A Combat Veteran Shares His Secrets for Success.

Introduction

How do you rebuild a meaningful life after war?

In *How to Win at Life*, retired Lieutenant Colonel John J. McBrearty—known as "Colonel Mack"—shares a deeply personal and inspiring roadmap for turning hardship into purpose.

Through raw honesty, faith, discipline, and the healing power of writing, McBrearty reveals how he transitioned from the intensity of combat to a mission of service at home. This book is more than a veteran's story—it is a guide for anyone seeking direction, hope, and fulfillment after life's toughest battles.

Told through an engaging interview format, *How to Win at Life* explores resilience, creativity, grief, leadership, and the pursuit of purpose.

Whether you are a veteran, writer, mentor, or someone struggling to find your next chapter in life, this book will remind you:

Your life still has meaning. Your story still matters.

Chapter 1
Prioritizing Your Time

Question

Colonel Mack, you've taken on many different roles in life, such as a retired U.S. Army Lieutenant Colonel who fought for the U.S., a volunteer golf coach who spends his free time guiding children interested in golf, a successful author who has published several books, and a husband, father, and grandpa who deeply loves his family. So, tell me, how do you manage to do all these meaningful things?

Answer

My main influence is my Lord and Savior, Jesus Christ. Family values: Having faced death and destruction firsthand, you quickly understand what truly matters in life.

I often ask the Lord, "Why did you spare me in Iraq when others perished or were injured?" I truly believe that God spared me from harm in Iraq so I could return home and share our story. I believe my writing is a calling, and a higher power is somehow involved. Writing for me is nearly effortless when you have that kind of motivation behind you.

Chapter 2
So Many Genres, So Little Time

Question

You have publications in various fields and genres, such as the book *Children's Golf*, the children's book *Claire the Magical Dog*, three how-to books about writing and publishing on Amazon, a few military memoir books, and more; so, could you tell me why you chose to diversify your writing instead of sticking to one type of literature or genre?

Answer

I believe I have a lot to say. One thing I learned through my research on writing and publishing is that a noted author recommends trusting your emotions when approaching writing. That's what I do. I find what moves me emotionally, and I follow that feeling. I believe you won't fail if you follow your heart. Success doesn't always have to be financial; it can be emotional, such as experiencing a sense of fulfillment.

Chapter 3
Favorite Literature

Question

Out of all the types of literature, which is your favorite and why?

Answer

I follow my heart and passion. When I publish a book, the sense of accomplishment and pride is similar to receiving an Academy Award or attending your college graduation. It is overwhelming and wonderful. Generally, after completing a book, I feel most passionate about that particular project. So, my favorites change depending on whatever project I am working on. In short, each genre of my writing is my favorite at different times.

Chapter 4
Favorite Book

Question

Among the books you've written, which one is your favorite and why?

Answer

That is a very difficult and complicated question. It's like asking which of your children is your favorite and why. The book I probably feel closest to is my simplest one, my children's book, *Claire the Magical Dog*. It features my entire family as characters, including the magical Shih Tzu dog. I've written 51 stories for that series, with 15 ready for illustration. I put a lot of thought into those books because I genuinely want to leave a positive legacy. Much effort is spent on developing meaningful life lessons in each story, all the while entertaining children and parents.

e

Chapter 5
From Writing to Publishing

Question

What inspired you to go from simply loving writing to publishing books?

Answer

As I told Amy Johnson of CBS's Veterans Voices telecast, I did not want to spend two or three years negotiating with a publisher over a single book. There's so much red tape with literary agents and publishers; I don't have time to mess around. Am I giving up the chance at a great fortune? Probably, but that's never been a motivation for me. So, by the grace of God, I found Written by Veterans, the veteran writing support group I mentioned earlier. They are sponsored by California State San Bernardino (CUSUB). Their founder, Andreas Kossak, an ex-adjunct professor from CSUSB, took me under his wing and mentored me with my writing and self-publishing goals. I am very grateful to him and our writing support group.

Chapter 6
Writer's Block

Question

What challenges have you faced while writing? For instance, writer's block, becoming emotional when writing about certain topics, hesitating to share your true opinions on specific subjects, or procrastinating? And how have you managed to overcome these challenges?

Answer

As I discuss in my *How to Get Published on Amazon* book series, I do not believe in writer's block. Too many aspiring writers use that as a crutch. I believe there can be brief pauses in the creative process, but not writer's block!

Emotions. When I wrote the *Citizen Soldier* book series, I had to relive a very dark period in my life. I lost my two parents due to natural causes and dealt with issues involving a borderline psychotic sibling, all while deploying to Iraq. Revisiting the combat experiences often brought tears to my eyes. However, this is where the therapeutic benefits of writing come into play; it was cathartic for me. By reliving those moments, I realized I had many bottled-up emotions and baggage. Writing my story helped me confront these issues and finally put them to rest. The result is a more well-rounded, peaceful individual, someone who now just wants to give back to his community, especially the veteran community.

Chapter 7
Inspiration

Question

What type of literature do you enjoy reading?

Answer

These days, I am very selective about what I invest my time in reading. I try to make the most of the time I have here on earth, so I don't waste it reading literature that doesn't interest me. I mostly read non-fiction. Sometimes, I read what other writers in my various genres are doing. But my favorite books to read—and I read a lot of them—are books about writing and self-publishing. I carry my book bag everywhere I go, and it always contains those books, along with highlighters, post-it notes, and other essentials. I often reread these books five, six, or even ten times. By the way, I also enjoy reading and listening to my own books as audiobooks.

Chapter 8
Writers' Support

Question

What kind of support do you receive from family, friends, or organizations for your writing pursuits?

Answer

One thing about writing: I find that being brutally honest with your work is the only way I can write. So, I will be honest with you on this subject, too. I do not get the support from family and friends I expected. I do not know why; maybe they have busy schedules?

I am part of ten writing support groups and even founded one of my own, which includes retired military general officers and police chiefs. I receive writing support from them and other groups, especially Written by Veterans. Other organizations, such as the Veterans Breakfast Club, the Military Writers Society of America, the Diamond Valley Writers Guild, and the California Writers Club of the Inland Empire, have published my work. I am very grateful for their support.

Chapter 9
Other Works

Question

Are there any upcoming publications you'd like to share?

Answer

Driving Range 101 will be released soon, followed by the autobiography My Military Life. The third book in the Claire the Magical Dog series, Claire Digs to China, will also come out this year. Several short motivational books will also be published.

Chapter 10
Awards

Question

I heard you recently won the National Veterans Creative Arts Festival and a scholarship. Congratulations! Can you tell us more about this contest and the scholarship?

Answer

The Department of Veterans Affairs hosted the competition, with 116 V.A. facilities participating. About 6,400 applicants competed nationwide. The categories included performance, artwork, and various types of writing. It was very humbling to place first in creative writing. I was also awarded a Therapeutic Arts Scholarship. I attended a week-long, fully funded National Veterans Creative Arts Festival held in Indianapolis, Indiana. The festival featured many writing workshops and an awards ceremony where my literary work was be displayed.

Chapter 11
Advice for Other Veterans

Question

Do you have any advice or suggestions for other veterans interested in writing or publishing?

Answer

Follow your heart and never accept no as an answer. Find something that moves you emotionally and write about it. Maybe start with a few daily notes in a journal. My *How to Write and Get Published on Amazon* books can give anyone a head start on how to begin writing.

Conclusion

The journey of life is not measured only by the battles we endure, but by what we choose to build after those battles are over.

This book offers more than just my perspective on writing, publishing, and perseverance—it reveals the deeper truth that purpose can be found even in the aftermath of pain. Like many veterans, I returned home carrying memories, burdens, and questions that could not be easily answered. The transition from military service to civilian life was not simple, and at times, it felt like the hardest mission I had ever faced.

But through faith, family, community, and the healing power of storytelling, I discovered something profound: **our lives are not defined by what we have survived, but by what we do with the life we have been given.**

Writing has become more than a hobby for me. It became a form of service. A way to honor fallen comrades. A way to confront grief and transform it into growth. A way to remind others, especially those fighting silent battles, that hope is real, and that no one walks alone.

If there is one message I leave you with, it is this:

Never underestimate the strength within you.
Never accept that your best days are behind you.
And never stop searching for the purpose that gives your life meaning.

Whether you are a veteran, a writer, a leader, or simply someone striving to live a better life, I encourage you to keep going. Your story matters. Your voice matters. And your next chapter may become the very thing that inspires someone else to survive, heal, and rise.

Thank you for allowing me to share this part of my journey with you. I hope you will continue walking with me through the pages of my future writings, as I explore faith, resilience, leadership, history, and the enduring spirit of those who refuse to quit.

The mission is not over. In many ways, **it is only just beginning.**

Every Veteran Has a Story.
What's Yours?

Since 2008

What is next?

Veteran's Breakfast Club Interview with Combat Veteran John J. McBrearty

The mission of the Veterans Breakfast Club is to build communities where people listen to veterans and their stories, ensuring this living history is preserved. They believe that through these efforts, people will feel connected, informed, healed, and inspired.

One such inspiring veteran is John J. McBrearty. In this publication, John, along with the Veterans Breakfast Club members, shares his three-decade-long military career and where that path has now taken him.

From John's initial military involvement as a U.S. Naval Sea Cadet at age 15, to U.S. Marine Corps boot camp at Parris Island, South Carolina, then to officer training at Quantico, Virginia, followed by active duty and reserves with the U.S.M.C., and eventually service in the National Guard, John shares the exciting journey that took him around the world. In conclusion, this publication explores leadership and humility, aiming to guide readers toward self-reflection and personal growth.

VETERANS' SURVIVAL GUIDE

PTSD is Not a Four-Letter Word!

Volume 1

by John J. McBrearty

Foreword by

Major General Peter J. Gravett

U.S. Army (Retired)

Appendix 1

VETERANS' SURVIVAL GUIDE

PTSD is Not a Four-Letter Word!

A podcast inspired the idea for this book with Chief Aaron Q. Seibert and Lieutenant Colonel John J. McBrearty. The broadcast was profound, drawing hundreds of veterans. Whether it was Chief Seibert's skillful delivery of his questions or Lieutenant Colonel McBrearty's responses, there was something their other veterans needed to hear.

Both Seibert and McBrearty are disabled combat veterans who, without knowing it, crossed paths in a veteran's support group. Perhaps fate brought these two veterans together, as both have dedicated their post-military lives to helping other veterans.

This book stands as a testament to those claims. You cannot finish reading it without feeling pride for our veterans and a deeper love of country.

Watch how these two veterans touch the souls of their brothers and sisters in arms as you've never seen before.

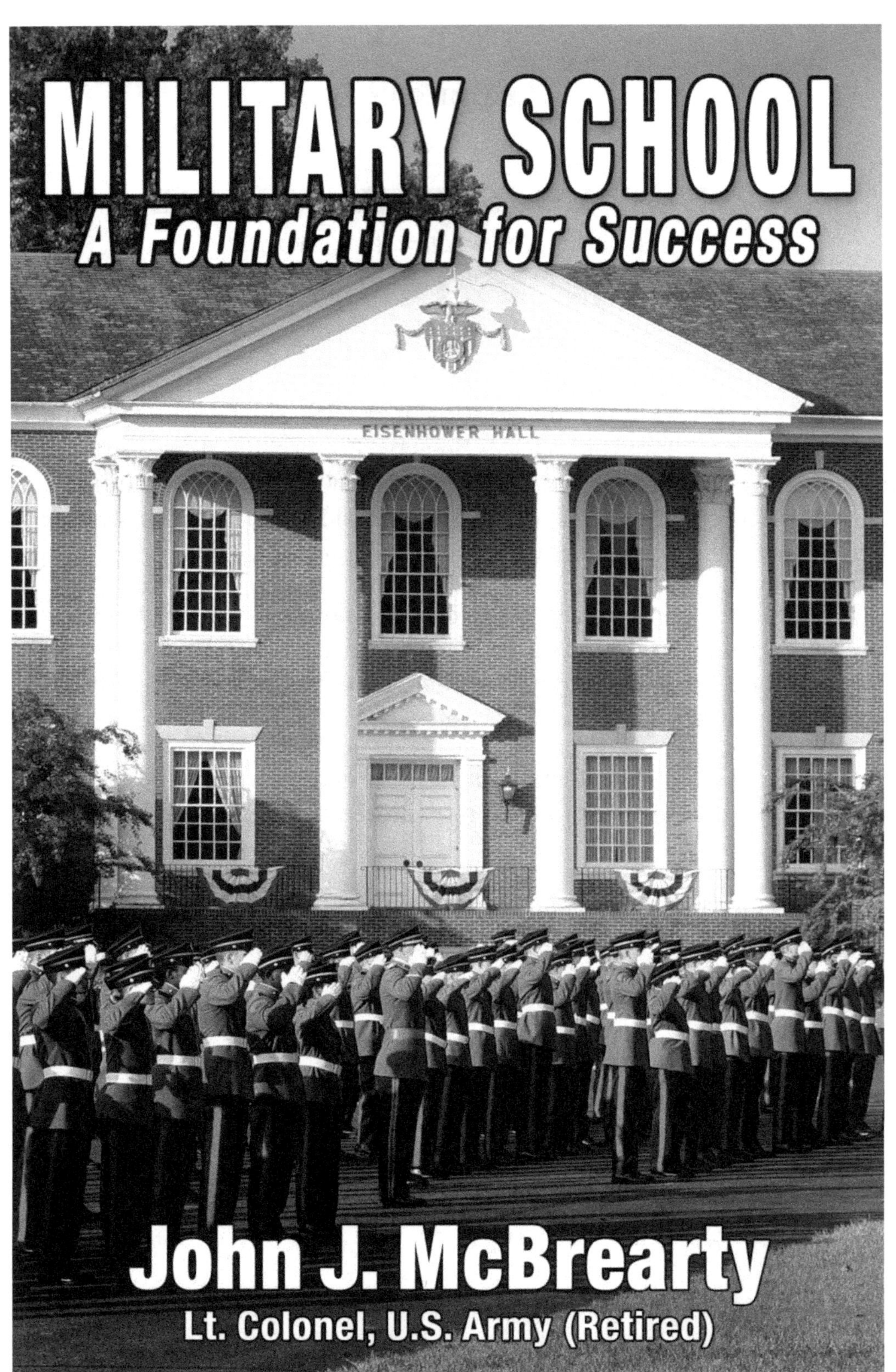
MILITARY SCHOOL
A Foundation for Success
EISENHOWER HALL
John J. McBrearty
Lt. Colonel, U.S. Army (Retired)

Appendix 2

MILITARY SCHOOL

A Foundation for Success

What did going to military school do for me?

By attending military school, I received not just an education; the military lifestyle at these academies enhanced the experience. These institutions not only offer a solid college-preparatory education but also emphasize character development, leadership training, and a rigorous academic routine. These virtues benefit not only future military leaders but also upcoming corporate managers. Traditional values form the core of their curriculum. The structured environment promotes self-improvement and academic success. I attended Valley Forge Military Academy and College (V.F.), and most examples in this publication come from my personal experiences there. However, I believe the other military schools listed in Appendix 1 offer a similar experience to mine at V.F.

Find out how attending *Military School* laid the foundation for my success.

HOW TO GET PUBLISHED ON AMAZON

Volume 1

"A Guide for Beginning Writers"

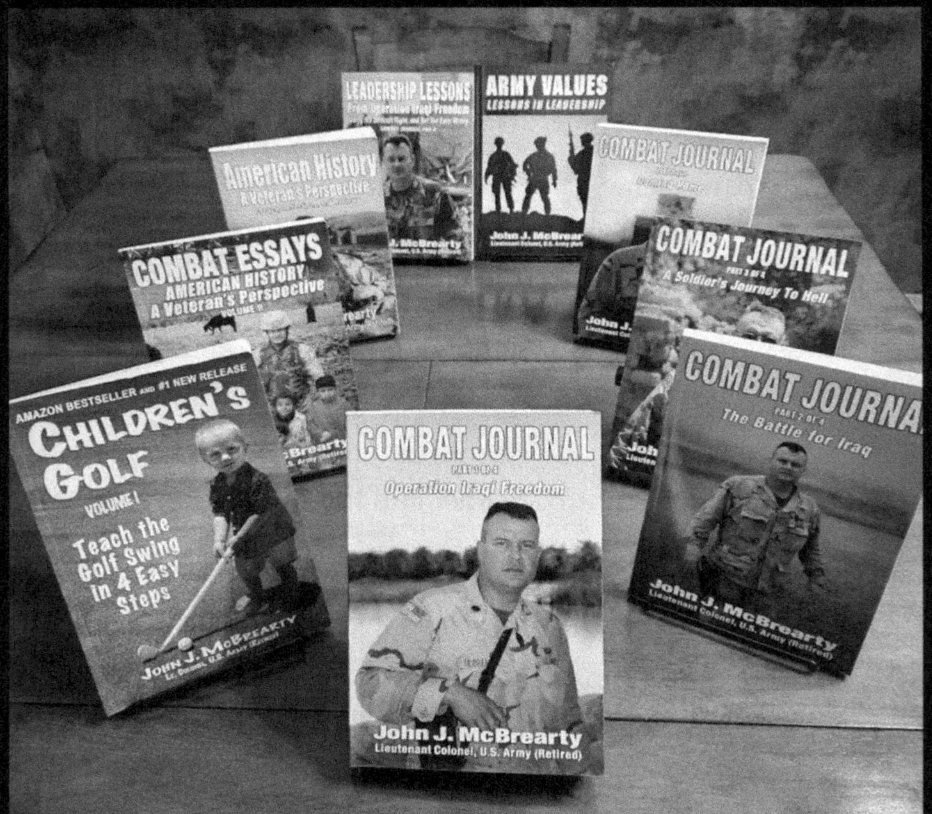

John J. McBrearty

Lieutenant Colonel, U.S. Army (Retired)

Appendix 3

HOW TO GET PUBLISHED ON AMAZON

A Guide for Beginning Writers

How did I start writing? Better yet, what makes me an authority on the subject? Perhaps I am some sort of literary genius? My answer to that: heavens no! I am a retired military officer with extensive formal education, but trust me, I am no genius. In fact, here is an Amazon 5-Star review on this very topic.

"John is no genius; however, his writings have moments of brilliance!"

One thing I have learned from a lifetime of leading Marines and soldiers is the KISS principle (Keep It Simple Silly – sometimes called Keep It Simple Stupid). I applied this principle when thinking about writing and publishing. Just like I did with my *Children's Golf* book, I broke the golf swing down to the simplest of terms so that a mere child could understand. I took the same approach with this book. I am sharing with you what I did on the road to writing and, eventually, publishing successful manuscripts.

What will you take away from this book?

If you read, study, and follow my methods, you will develop into a writer. All it will take is a little effort on your part. Through the use of my simple yet effective writing exercises, you can be up and running as a writer in no time.

John J. McBrearty
Ten-time published author.

"John is a prolific author writing about American history and golf books for children."

Jim Hill, CBS Los Angeles, *Sports Central*

HOW TO GET PUBLISHED ON AMAZON

Volume 2

From Publishing Zero to Publishing Hero!

John J. McBrearty

Lieutenant Colonel, U.S. Army (Retired)

Appendix 4

HOW TO GET PUBLISHED ON AMAZON

Everything you need to know about how I went from a publishing zero to a publishing hero within one year!

How do you publish a book? That is the $10 million question, and we might actually have the $10 million answer right here in these pages.

What will you take away from this book?

My theories have a proven track record. I want to share that knowledge with you. In this book, I discuss my journey of writing and publishing my first, second, and third books, and beyond. My simple yet effective secrets to literary success are explained in detail in the following pages. If you follow my easy-to-understand methods, you too can go from a publishing zero to a publishing hero in no time.

John J. McBrearty

Veteran/Author/Father & Husband

"Now retired from the military, COL Mack has published eleven books and became an Amazon Best Seller. He pays tribute to those he served with."

Chris Holmstrom, CBS News, Los Angeles *Veterans' Voices*

HOW TO GET PUBLISHED ON AMAZON

Volume 3

Inspirational Quotations and Other Writings by a Combat Veteran

John J. McBrearty

Lieutenant Colonel, U.S. Army (Retired)

Appendix 5

HOW TO GET PUBLISHED ON AMAZON

Inspirational Quotations and Other Writings by a Combat Veteran

If you are looking for inspiration, this book is for you!

In the following pages, Lieutenant Colonel John J. McBrearty, U.S. Army (Retired), explores various literary principles and provides examples. These examples include his own original inspirational quotes, book summaries, reviews, and blurbs. This literature is unique because it comes from a retired military officer with worldly insights that few of us could ever imagine or achieve. John's candid and clever expression will likely inspire you.

For aspiring authors looking to get published on Amazon, the following pages provide useful examples of what can help an author succeed.

"John is a veteran who became an Amazon Best Seller for his books on American history.... his words are inspirational...."

Amy Johnson, CBS Los Angeles, *Veterans' Voices*

Appendix 6

Literary Works of John J. McBrearty

Leadership Books

LEADERSHIP LESSONS (1), *From Operation Iraqi Freedom*
ARMY VALUES (2), *Lessons in Leadership*

American Military History and Memoir Books

AMERICAN HISTORY, A VETERAN'S PERSPECTIVE (1), *Essays, Reflections, and Reviews*
COMBAT ESSAYS (2)
MY MILITARY LIFE (3), *From Sea Cadet to Lieutenant Colonel, U.S. Army, NO REGRETS!*

COMBAT GOLF: *An Army Major Builds a Golf Course in the Iraq War*

Combat Journal Book Series

FROM HOME TO HELL (1)
FIRST DAY, FIRST FIREFIGHT (2)
ON PATROL IN IRAQ (3)
THE BURDEN OF COMMAND (4)
WHISPERS OF MUTINY (5)
IRAQ TANK BATTLE (6) (Bonus Edition)

CITIZEN SOLDIER: OPERATION IRAQI FREEDOM-A MEMOIR (Compilation Edition)

How to Write and Get Published Books

LITERARY WORKS OF JOHN J. MCBREARTY
HOW TO GET PUBLISHED ON AMAZON
Volume 1, *A Guide for Beginning Writers*
Volume 2, *From Publishing Zero to Publishing Hero!*
Volume 3, *Inspirational Quotations and Other Writings by a Combat Veteran*
Volume 4, *Podcast 101: A Must for Any Author*

Sports Books
How to Golf Book Series

CHILDREN'S GOLF **(1)**
DRIVING RANGE 101 (2)
GOLF COURSE 101 (3)
ADVANCED GOLF TECHNIQUES ANYONE CAN LEARN (4)

Children's Books

Claire the Magical Dog Book Series

CLAIRE MEETS DEAN AND JACOB (1)
CLAIRE INTRODUCES BABY TO GOLF (2)
CLAIRE DIGS TO CHINA (3)
CLAIRE CELEBRATES VETERANS DAY (4)
CLAIRE AND THE SERVICE DOG (5)

Motivational Books

How to Win at Life Book Series

VETERANS' SURVIVAL GUIDE (1), *PTSD Is Not a Four-Letter Word!*

MILITARY SCHOOL (2), *a Foundation for Success*
HOW TO WIN AT LIFE (3) *A Combat Veteran Shares His Secrets for Success*

Anthologies

ANTHOLOGY 5, *Written by Veterans, CSUSB*
MESSAGES FROM THE BACKROOM, *V.A., Loma Linda, CA*

Author John J. McBrearty.
(Photo is courtesy of Nicolle Rea)

About the Author

Lieutenant Colonel John J. McBrearty, U.S. Army (Retired), served honorably in our nation's military for 32 years. His career spanned the globe, with operations in Iraq, Kuwait, Japan, Australia, Mexico, and Thailand.

John J. McBrearty, a member of Phi Beta Kappa, is a magna cum laude graduate of Valley Forge Military Academy and College as well as Temple University. He also holds a master's degree in American history from American Public University. He served as an Assistant Professor of Military Science at California State University, San Bernardino, and Claremont McKenna College.

John J. McBrearty's literary works have been published or featured *in Sports Illustrated, Golfweek, Golf Digest, Stars and Stripes, Callaway Magazine, Grizzly Magazine, the Los Angeles Times, the Orange County Register, the Main Line Times, the News of Delaware County, the Daily Bulletin, the San Bernardino Sun, the Press-Enterprise, Veterans Breakfast Club Magazine, Military Writers Society of America Dispatches Magazine*, and others.

To date, John has published over twenty books. John also won first place in creative writing at the National Veterans Creative Arts Festival, earning him a Veterans Therapeutic Arts Scholarship.

Upon his retirement from public service, John was recognized for his excellence by President Barack Obama.

"Coach John" is pictured here at a book signing during a local charity golf tournament. (Photo is courtesy of MBM Publishing.)

Special Thanks

I want to thank my family: Lynette (my wife), Kristina (my daughter), John Jr. (my son), Dean, and Jacob (my grandsons).

I also want to thank all the soldiers and Marines I served with over the years, especially the five soldiers from my unit who lost their lives in combat.

Thank you to my literary mentors, **Andreas Kossak**, Valerie Ormond, Cindy Rinne, John Cole, Peter Schreiber, Derrick Dardin, Ph.D., Owen McCue, Peter J. Gravett, Kelly Galvin, and the other members of my writing support groups.

To the Will family: Nancy, Steve, Derek, and Chelsea, thank you for your loyalty and ongoing support.

To my technical support team: Nichole Rey (photographer), Mike Waitz (proofreader), Christian Resendez (media), and Rey Soto (media and illustration).

To the Department of Veterans Affairs and its exceptional employees.

To my Lord and Savior.

God bless each of you.

John J. McBrearty

LEGACY

Writing is like piloting an aircraft; it is a perishable skill.

Like writing, if you don't put the time into the simulator, you will be ineffective in the cockpit.

John J. McBrearty
September 11, 2025

Follow John J. McBrearty's other literary works at:

https://linktr.ee/johnwriteshistory

1-800-273-8255 PRESS 1

NEW NUMBER,

SAME SUPPORT.

Dial 988 then Press 1.

Share it with your networks.

www.ingramcontent.com/pod-product-compliance
Lightning Source LLC
LaVergne TN
LVHW081321110826
845149LV00006B/1559